THEORIES OF THE UNIVERSE

And existential philosophy

2020

ETHAN RUEDLINGER

© Ethan Ruedlinger

The purpose of this book is to shed light on some of the more obscure existential philosophy and theories of the origins of the universe, as well as the existence of the multiverse, time, and higher planes of existence. My hope is that this book will ignite fervent discussion among the youth of the astronomy community, and inspire new generations to explore and discover the final frontier, space!

This book is dedicated to my father in law, who shares my passion for the mysteries of the universe.

INDEX

Chapter 1: The Origin of the Universe. . .   *1*

Chapter 2: The End of the Universe. . .   *15*

Chapter 3: The Atom-Universe Theory. . .   *24*

Chapter 4: Multiversal Travel. . .   *28*

Chapter 5: Creating Matter From Nothing? . .   *37*

Chapter 6: Time Travel. . .   *46*

Chapter 7: Full-Dive Virtual Reality. . .   *55*

PROLOGUE

Since the primordial era of man's existence, we have been captivated by the stars in the night sky. Mankind has sought out answers to the inner workings of the cosmos for millennia. Existential philosophers, Astrophysicists, and even amateur astronomers all playing key rolls in the discovery of the laws of nature and the universe. From Galileo Galilei, to Einstein, to Neil deGrasse Tyson, the world's greatest minds have made astounding discoveries and advancements in the field of science. I have written this book on some of the most interesting theories of science, , including versions of the multiverse theory, interdimensional travel, time travel, teleportation, and theories of what our universe could constitute beyond our current perception.

CHAPTER 1

THE ORIGIN OF THE UNIVERSE

"If the universe has a beginning, it's beginning, by the very condition of the cases, was supernatural; the laws of nature cannot account for their own origin."

-John Stuart Mill

Currently, the most widely accepted theory of the origin of the universe is the "Big Bang Theory."

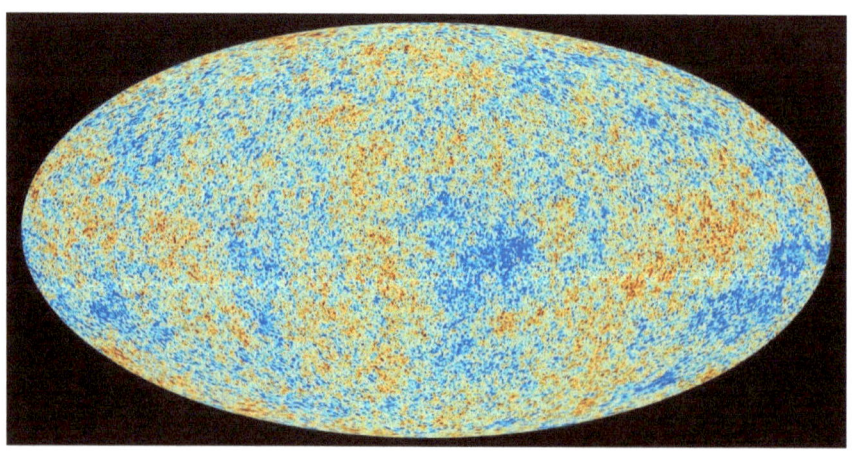

To put it simply, this theory states that everything in the known universe originated from a single one-dimensional point in space. An infinitely dense ball of mass known as a singularity. This singularity violently exploded outward, creating a hot pool of neutrons, protons, anti-electrons, electrons, photons, and neutrinos.

After three minutes, light elements were made through a process called: "nucleosynthesis."

Within a few minutes, as the universe continued to expand outward, the raging hot pool of subatomic particles began to cool, allowing the formation of atomic nuclei, mainly hydrogen and helium, the two main elements involved in the process of nuclear fusion.

As the universe cooled and rapidly expanded outward, stars began to form. Protoplanetary disks began to form around stars, giving rise to planets, and eventually, the formation of Earth.

This theory is widely accepted among the scientific community, and is widely regarded as the true origin of our universe.

However, it's not the only theory as to the origin of the universe.

Another prominent theory, known as the Steady-State theory, supported by the legendary Stephen Hawking and originally proposed by Sir Fred Holye, claims that our universe had no beginning. In other words, it's the idea that the universe we live in is not only infinitely vast, but has an indefinite age, an eternal universe. Now, let's imagine for a moment, that our universe is, in fact, an infinite expanse of galaxies, of equally dense proportions and homogeneity, it's nearly impossible to account for the fact that each celestial body is moving away from one another at a fixed rate of speed. Taking this into account, we can say with confidence that, unless some unforeseen force has magically appeared from thin air to repel celestial objects away from one another, our universe could not have existed eternally up to the present, thus ruling out this theory.

That's not to say, however, that our universe couldn't be an infinite expanse of galaxies, gas, and dust. Perhaps our universe is much older than we've anticipated, and therefore time has allowed the near-infinite expansion of our universe at a constant velocity.

This would mean our observable universe is merely an indication of the amount of time the light emanating from stars has had to reach us. Perhaps our supercluster is young, created recently on the outskirts of the ever-expanding universe.

Another theory, mostly denounced by the scientific community, is the hypothesis that all of existence as we know it is being simulated somewhere, by some other beings or entities, a higher form of intelligence. This theory in principle is nothing new, and has been likened to the Greek philosopher Plato's "The Allegory of the Cave." The story compares people who have no knowledge of the Theory of Forms to prisoners in a cave. They are chained, unable to turn their heads, all they can see is the wall. Behind them, there's a fire burning. And there are puppet masters behind them, who use their puppets to cast shadows onto the cave wall. The prisoners are only able to see the shadows the puppets produce, not the puppets themselves. Effectively, the prisoners would perceive the shadows as being the real deal, having no way of knowing or learning of the existence of the true puppets.

In the same way, our universe, our reality, and everything we perceive on a daily basis could be simulated, or at least, not the highest plane of reality. Our universe could be a quantum simulation, or an advanced computer simulation. The only things we know, or perceive, are things our brain tells us. Everything we've ever known is processed through our brain, which could be manipulated by some unseen intelligent force.

Think of it this way, if we put a bucket over your head, so no nerve impulses were able to reach your brain from your senses, and we send artificial neurological signals directly to your corresponding neurons, causing you to feel simulated sensations, we could effectively create an entire simulated world. The recent findings of quantum physics shed doubt on the "fact" that our universe is material. John Wheeler said that physics has evolved from the premise that "everything is a particle" to "everything is information." Henceforth, the definition of a particle is fuzzy, leading physicists to conclude they could be a qubit; a quantum computing bit. The one responsible for this hyper-realistic simulation we could potentially be in is described as a "Programmer in the next universe up." Famous tech entrepreneur Elon Musk famously said: "The odds that were *not* simulated beings are one in billions."

Future generations may possibly possess powerful "Mega-Computers" capable of running simulations of past scenarios, or in other words, simulating their ancestors. The simulated beings could be endowed with a sort of artificial consciousness, or a real consciousness that is being generated by artificial means. The way humans could achieve such complex simulations is by studying and understanding the inner workings of human consciousness, as well as fully mapping the function of each neuron in the brain, so they can be manipulated individually to produce false, detailed sensations experienced by the individual.

    Understanding human consciousness will also allow us to

create an artificial consciousness, or a man-made being.

While hypothetically more likely than the multiverse theory, most scientists rule this theory out as being nothing more than crackpot pseudoscience. My personal belief is known as Creationism. It's the belief that the Lord God created the heavens and the earth, as it is written in Genesis. The Lord created everything we see before us today, and sent the light of distant stars 13.8 billion light years forward, so we may have a beautifully lit night sky. The entire universe is much younger than current science predicts through carbon dating and astronomy.

Another theory is known as "Last Thursdayism."

To put it bluntly, it's the belief that the world was created last Thursday. Everything around you, the universe, the light from distant stars, history books, memories in your mind, everything was created to appear 13.8 billion years old, even though everything is only a week old. No matter what you say or do, you cannot disprove this hypothesis, as it is an Omphalos Hypothesis. This theory is similar in principle to most forms of creationism, in the idea that the world was created to appear much older than it truly is.

CHAPTER 2

THE END OF THE UNIVERSE

"But of that day and that hour knows no man, no, not the angels which are in heaven, neither the son, but the father."

-The Lord God, Mark 13:32

There are countless possibilities pertaining end of the universe, many of which entail either a peaceful or horrific end to our existence.

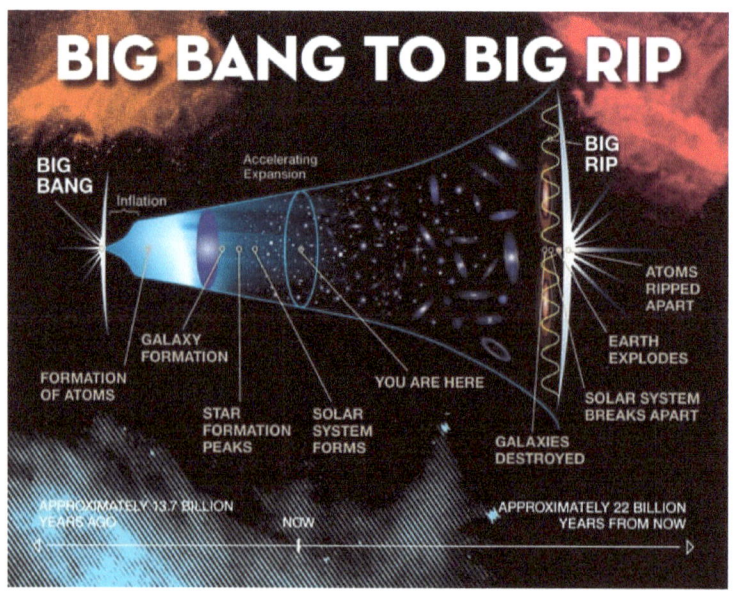

One theory of the ultimate fate of the universe is known as the "Big Rip."

Recently, scientists have calculated that the universe is expanding at a much faster rate than previously surmised, concluding that the rapid rate of this expansion may cause the universe to rip apart. The Big Rip, if it were to occur, would destroy the universe within 22 billion years from the present.

Scientists attribute the recent increase in the speed of the universe's expansion to the work of dark matter. In order for the Big

Rip to occur, dark matter must win its battle over ordinary matter, to the point that it can rip apart individual atoms.

Another theory concerning the end of the universe is known as the "Big Crunch." Rather than rapidly expanding indefinitely, if the universe were to reach a point where matter decreases over time, gravity could shift to the dominate force. This would cause the universe to shrink and collapse in on itself, perhaps even creating a new singularity. Researchers in Denmark believe they have found enough evidence to claim that this is already occurring, and that the process, known as "phase transition," is effectively 'eating away' at the cosmos. Scientists assure us, however, that the universe collapsing would be a "quick and painless death."

The third theory is known as the "Big Freeze."

The hypothesis commonly referred to as the "heat death of the universe," it is widely regarded as the most probable outcome for the fate our universe, in accordance with our current understanding of the laws of physics.

Entropy originated from a principle of thermodynamics covering energy, and refers to the idea that every system eventually moves from order to disorder. Entropy is the scientific measurement of the shift from order to disorder in systems. In the Big Freeze hypothesis, everything in the universe will reach a state of "maximum entropy," in which all heat in the system (the universe) will be evenly distributed across the cosmos, therefore there will be no room for usable heat or energy to exist, and the universe will die a "heat death." Time will become an endless void where nothing can occur, as there is no usable energy left in the universe.

There are more obscure theories of the end of the universe, such as the theory that the universe's expansion will slow to a halt, expanding at an almost unnoticeable rate.

A few theories involving dark matter, as related to the ultimate fate of the universe, involve dark energy as the main driving force of the universe.

One theory is that Dark Energy never weakens, causing the universe to infinitely expand, and every celestial body that's not a part of our local group to be pushed out of sight over time.

Another hypothesis is that dark energy grows more powerful over time. This would lead up to the occurrence of the "Big Freeze." There are more theories, such as the idea that dark energy could decline in strength over time, or even convert itself into normal matter and energy to replenish the universe and prevent "heat death," much like what occurred during cosmic inflation and prior to the hot Big Bang.

Formulas for determining

cutoffs tell us that the universe is in a state that would've required it to have began nearly 13.7 billion years ago, and will reach it's "cut-off" in 5 billion years. That means, according to popular opinion, the universe will cease to expand in 5 billion years. What this will mean for humanity and existence as a whole depends entirely on the forces at play, and the physics of the expanding universe.

I believe the universe will never end. My personal belief is that Jesus Christ will renew the Earth 1,000 years after the "Apocalypse" describes in revelation. After reigning with Christ in heaven for 1,000 years, we will return to Earth and live as Adam and Eve did, as the Lord God originally intended for mankind to live. The Earth will be eternal, just as Heaven is.

CHAPTER 3

THE ATOM-UNIVERSE THEORY

"I, a universe of atoms, an atom in the universe."

-Richard P. Feynman

A not-so-popular theory pertaining to what lies just beyond our universe, is the Atom-Universe theory.

As we know, everything in the universe is composed of matter, a

particle with a constitution of 99% "empty" space.

  Could it be possible that the entire universe as we know it composes an atom of a larger structure? Perhaps this "macroverse" could be a car, a human, a garage, a tree, a Star, a planet, or some alien structure we can't even conceive, and our entire universe is contained in a single atom composing that universe. Perhaps there are other universes within the atoms that constitute everything in our universe, and perhaps this is an infinite chain of an unfathomable amount of universes. This mind-blowing concept is really nothing new, and has been pondered for generations. However, the only way to prove it would be if future technological advancements permit interstellar travel, and we're able to escape this universe and view the larger structure we're all a part of from afar. In a certain sense, this would also prove the multiverse theory, as our neighboring universes would also be contained within atoms/atomic nuclei. The implications of this theory would have a profound impact on science, and would be labeled as the greatest discovery in human history. There are 10 to the 82nd power atoms in the known universe. That's an absolutely inconceivable number. Now, imagine if our universe is just one atom constituting a much larger universe. Forget the number of galaxies, the multiverse would be massive! There would be more universes in existence than grains of sand on Earth! By the sheer size of our own universe, we are insignificant. But if we exist in a multiverse, we're nothing compared to the rest of existence!

CHAPTER 4

MULTIVERSAL TRAVEL

"So multiverse or not, we still have to come to terms with the origin of the laws of nature. And the only viable explanation here is the divine Mind." -Antony Flew

Assuming multiple universes exist within our reality, will humanity ever conquer the monumental task of breaking down the barrier of distance between the infinite worlds through use of teleportation or some form of portal/wormhole-like technology? The answer is complicated, to say the least.

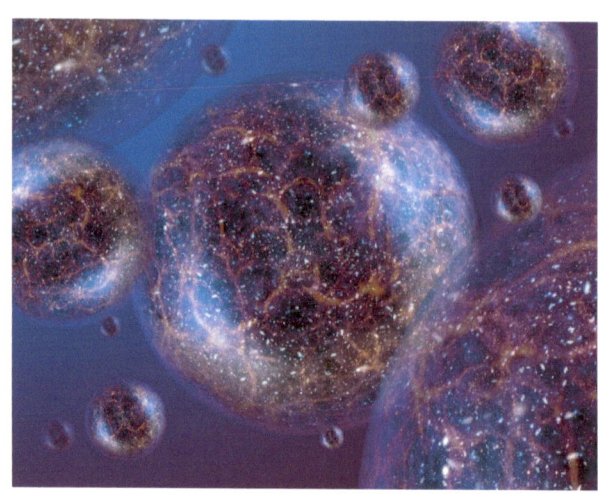

Our civilization is undergoing what scientists refer to as "The Great Filter." If you take into account the vastness of our own galaxy, along with our knowledge that intelligent life has come into being at least once, there should be evidence for alien life. But there isn't. This is known as the Fermi Paradox, names after physicist

29

Enrico Fermi, who examined the inherent contradiction between the mathematically calculated high probability of the existence of extraterrestrials, and the lack thereof. Fermi summed this up by asking "Where is everyone?" However, a more fitting question may be "What happened to everyone?"

Alien life is extremely likely, however, we've been searching for decades through radio telescopes, and even unmanned interplanetary missions, in which we've failed to discover evidence.

The most likely explanation for

this is "The Great Filter."

The idea is that, at some point along life's development, there is a massive challenge that destroys alien life before it becomes intelligent enough for us to discover— a great filter.

This filter could take on a variety of forms. It could be that the formation of planets in the "Goldilocks Zone," or the habitable zone, is extremely unlikely. Or it could be that the formation of prebiotic organic molecules capable of accumulating into life is unlikely.

We've estimated there to be over

40 billion planets in the Goldilocks zone in the Milky Way alone, but perhaps the conditions on these planets are not quite ideal for life.

The Great Filter could occur in the very early stages of life.

You're probably very familiar with the phrase: "The mitochondria is the powerhouse of the cell."

It is theorized that mitochondria were, at one point, separate living entities from bacteria. However, a signal celled organism attempted to consume a mitochondria, and rather than being digested, the bacterium teamed up with the cell, producing extra energy so the cell could develops into higher life forms.

An event similar to this may do so unlikely, that its only occurred once in our galaxy. Or, the great filter could be the development of large, intelligent brains. Perhaps other life forms in our galaxy are animalistic in nature, not possessing the intelligence needed to develop civilizations.

These ideas assume the great filter is behind us, that we've been lucky enough to have surpassed every possible downfall of life. The discovery of nuclear power is a probably way a society could destroy itself. Using a planet's resources to develop an advanced civilization can also destroy the planet, leading to mass extinction. A good example is air and water pollution.

Or, it could be an entirely unknowable disaster that we're unable to foresee at our current level of technology.

So, if we can surpass The Great Filter, and our technology advances over a period of several million, or maybe even billions of years, we will most likely be able to travel between universes. Although the laws of physics would most likely vary between different universes, there may still be some sort of common medium through which we would be able to travel to other universes. If so, we will be able to expand our civilization infinitely across an infinite number of stars, galaxies, and other planes of existence. It's not at all far-fetched to believe that eventually, we will have mastered the laws of the universe, and gain the ability to bend them at will. With an extremely advanced civilization, the doors technology could open are limitless.

CHAPTER 5

CREATING MATTER FROM NOTHING?

"The more I study science, the more I believe in God."
-Albert Einstein

The supposition that humanity will develop the ability to create matter from nothing seems preposterous in first glance.

However, as previously mentioned, assuming humanity surpasses the great filter, and progresses towards becoming a Type IV or Type V civilization on the Kardashev Scale, it is sensible to presume humanity will achieve the necessary level of technology to create matter and energy from absolutely nothing.

The Kardashev Scale is a method of measuring the level of technological advancement a civilization has achieved.

A Type 0 Civilization, which is humanity's current level of advancement, would not be ranked on the Kardashev Scale. In 1973, Carl Sagan estimated humanity to be a 0.7 on the scale. Some theoretical physicists, such as Michel Kaku, estimate humanity as being 100-200 years off from being promoted to a Type I civilization, and perhaps billions of years from becoming a Type IV or Type V civilization.

A Type I civilization, also known as a planetary civilization, can utilize and store every bit of energy available on it's planet. This means utilizing the energy of the sun to meet the energy demands of an ever-growing population. We would need to increase our current energy production over 100,000 times to achieve this, as well as possibly find a source of energy for the human body apart from the consumption of animals and plants.

A Type II civilization, also known as a stellar civilization, can utilize and store every bit of the available energy of it's host star. There are many theoretical ways of achieving this, including a "Dyson Sphere." This device would encompass every inch of the star, gathering all of the energy the star emits and storing it for use.

A Type III civilization, also known as a galactic civilization, can utilize and store every bit of the available energy of it's host galaxy. Feeding each habited planet with the energy emissions of millions of stars, the civilization would have no problem achieving things such as light speed travel and teleportation.

A Type IV civilization, also known as a universal civilization, can utilize and store every bit of energy within it's universe. This type of civilization would need to utilize forms of energy generation that are completely unknown and unfashionable to us, these beings would be masters of the universe, with the ability to move, create, and destroy entire galaxies at will, as well as traverse the ever-expanding universe. These beings may also have the ability to unlock the secrets of black holes.

A Type V civilization can utilize and control every bit of energy within multiple universes.

Assuming a Type IV civilization has mastered the physical forces of it's host universe, dark matter and dark energy could theoretically be harnessed as "extragalactic energy" by humans to generate matter from nothing. Based on our current understanding of the laws of physics, as Einstein's equation E=MC2 suggests, matter such as oil and coal can be converted into energy such as electricity, and energy such as photons can be converted into matter such as subatomic particles, but neither can be created from nothing. However, it is possible that humanity, upon reaching a Type V civilization, could unlock the secrets of bending the laws of the universe by studying other universes, or by simply using the differentiating laws of other universes that could potentially allow for the creation of matter and energy.

   Such a civilization could create

their own custom universes, perhaps even their own microverses. This is entirely possible, if God decides to allow humans to continue their existence for a period of billions of years, and continues to encourage our development on technology, we could control and create entire universes as a species. This would require us to extinguish war, come closer to God, take better care of our planet, and support scientific advancements.

Perhaps God's intention is to encourage us to build our civilization up to a Type V after the events of Revelation, the 1,000 years of peace, and the believer's return to Earth. Perhaps God will allow us to develop our societies into a Type V civilization capable of creating infinite universes.

CHAPTER 6

TIME TRAVEL

"Once confined to fantasy and science fiction, time travel is now simply an engineering problem." -Michio Kaku

Oddly enough, one of the biggest issues with time travel is the existence of time itself.

Time may only be a fabrication of the human mind. Time itself may not exist in a physical form of which can be manipulated, but rather, may not exist at all. Time as we perceive it could be created by the human mind, when in reality, "time" marches on at the same pace everywhere in the universe, as an untouchable, nonexistent entity.

Einstein's theory of relativity predicts the existence of black holes, white holes, wormholes, time dilation, and many other such phenomena. The existence of black holes has since been proven, and photographed in 2019, as seen here:

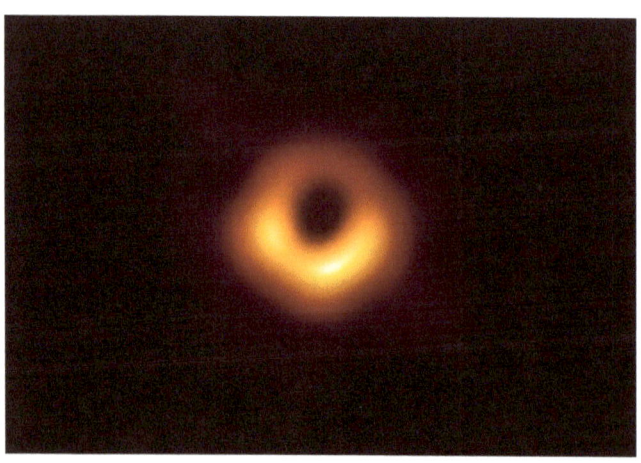

The first ever image of a supermassive black hole in the center of Messier 87 and it's shadow.

This is the first-ever direct evidence of a black hole, supporting a theory that Einstein mathematically calculated nearly a century prior. The discovery was named the "Scientific Breakthrough of 2019."

Although we've proven a portion of Einstein's theory, who's to say he couldn't have gotten one little thing wrong.
Einstein predicted that the gravitational pull of celestial bodies would warp space-time. In other words, if you're near a black hole, given their extremely

powerful gravity, time will go by much slower for you in comparison to how time passes on earth. One day for you could be several years on earth. The same is true for traveling at the speed of light. If you were to travel close to or at the speed of light, you could theoretically stop time around you, time is only moving for those who are on earth, while, to those who are onboard the ship, the time it took to travel to their destination passed instantly. Using this knowledge, one could estimate that you could travel at the speed of light and arrive millions, or even billions of years in the future, like a makeshift time machine. This is what's known as time dilation.

This is one of the ways scientists claim we have already traveled in time.

Let's say an astronaut lives on the moon for 82 years. When he returns to earth, he will be a millisecond younger than the rest of the world. However, scientists debate on what causes this. Some would argue that clocks operated differently and at different rates of speed relative to the force of gravity being exerted on them, explaining the time difference as nothing more than the unreliability of clocks and watches in space.

The majority of scientists argue in favor of the theory of relativity, stating that time dilation is a legitimate physical process that occurs with the excessive assertion of gravitational force on the fabric of space time that warps how fast, or how slowly, time passes relative to other portions of the universe.

Time is known as the fourth dimension. Forward and backward being the first dimension, left to right being the second dimension, and up and down being the third dimension. But, we can't simply move through time at will, similar to how the classic two dimensional Mario can't move forward or backward relative to us, , only side to side. If Mario were to be a living two dimensional entity, we could observe him in his two dimensional world from our third dimensional world with him being none the wiser. In the same way, it has been speculated that the reason why "fourth dimensional" beings, or, time travelers, haven't visited us. Well, perhaps they have. They could be watching you read this book, from the safety of their fourth dimension. We wouldn't have the ability to observe them, as we are mere three dimensional beings.

Using this principle, we could conclude that the possibility of an infinite number of dimensions, and their perspective inhabitants, is not only possible, but perhaps even probable. A higher dimension would be equivalent to a higher plane of existence, considering a two dimensional being would be inferior to us three dimensional beings. This could also be the reason extraterrestrial life has decided not to contact us, or have ignored us entirely. A 10th dimensional being would view us as nothing more than an inferior, puny nuisance to their kind.

There are many hurdles in physics and our basic understanding of existence as a whole that we must overcome before a conclusion can be reached regarding the feasibility of time travel. For the time being, it will remain not only theoretical, but a hotly debated topic among the scientific community.

CHAPTER 7

FULL-DIVE VIRTUAL REALITY

"Virtual reality is the first step in a grand adventure into the landscape of imagination." -Frank Biocca

Full-dive virtual reality is perhaps one of the most interesting concepts of the 21st century.

Imagine we put a bucket over your head, cutting off every one of your senses. Now, imagine we start sending artificial signals to your brain from a computer that simulate experiences, and your brain sends signals to the computer in reaction. The result would be a virtual reality game indistinguishable from reality. To you, you would really be fighting monsters, in call of duty, in Mario, or any of your favorite video games! This sounds fantastic, right? It is possible, we've already advanced computer-to-brain interface technology to the point where we can scan neurological signals and recreate the pictures your brain sees, as well as recreating dreams. We are also able to implant a device into the brain that restores hearing. However, full-dive VR that encompasses the simulation of all human senses is much farther off from the technology of today.

According to the scientists who are currently working on this technology, there are three ways this can be achieved. The first method, the "Invasive" method, involves implanting electrodes into the human brain to simulate experiences. The experimentation require to achieve this is considered unethical by the US government, and therefore will not be permitted any time soon. The second method, the "Semi-Invasive" method, requires scientists to place electrodes onto the brain's surface. This is also considered unethical by the US government.

The third being non-invasive, which is the current method being utilized by scientists to ensure the ethical research required for the invention of full-dive VR. A map of the neurons in the brain, as well as their function, is slowly being completed through multiple EEG scans.

In order for full dive VR to work, you would have to deactivate the muscles entirely. This would be difficult to achieve, and there is no known way of doing this to date.

    The research for full dive VR in the area of neuroscience progresses slowly due to the government's ethical concerns over invasive and semi-invasive procedures in the brain. Non-invasive research is the only form currently permitted to be conducted by scientists in pursuit of full dive VR. Using an EEG, scientists perform "black box" testing to measure the brain's response to stimuli, giving us a map of the functions of neurons in the brain. However, our current lack of a complete map hinders our ability to utilize all human senses in full dive VR.

The hardware and computational power required to decode, process, and transmit such large amounts of data can only be achieved through quantum computing. EEG helmets are bulky, requiring 11-16 electrodes that have to be wet before being mounted onto the wearer's head. The helmets must also be attuned to each individual, falling short of the simplicity of plugging in a PlayStation and placing a helmet atop your head. There are many pioneers in the VR industry working to advance both "traditional" and "full-dive" VR technologies. As "Computer-to-brain interface technology advances, so does VR technology.

ABOUT THE AUTHOR

Ethan Ruedlinger is a US Army veteran, Shito-Ryu Kyokushin Karate black belt, and entrepreneur who possesses a deep fascination with all things history. Since his parents took him across the country visiting historical sites and museums during his childhood, he has a profound love for America and American history. Serving honorably in the US Army for a period of two years as a signal support systems specialist, Ethan has firsthand knowledge of the modern military. Ethan is also knowledgeable on astronomy, existential philosophy, and Christianity.

www.ingramcontent.com/pod-product-compliance
Lightning Source LLC
Chambersburg PA
CBHW040324220526
45473CB00009B/2554